First published in 2026
By Kids4kids.com.au publications
Giving a voice to kids to speak to tell their stories and speak to peers everywhere

A catalogue record for this book is available at the National Library of Australia

ISBN (Paper back): 978-1-7644009-3-0
ISBN (eBook – Amazon.com): 978-1-7644009-2-3

We are stardust
The Story of Stars for Curious Minds
By Anton Mohajeri

CHAPTER 1.
WHAT IS A STAR?

WHAT EVEN IS A STAR?

Every night sky is a giant storybook written in shimmering dots.

Each twinkle is a message from far away — a glowing postcard sent across space and time.

A star isn't fire, even though it looks like the world's biggest campfire.

Fire needs oxygen, and space has about as much oxygen as a brick.

Stars glow because they are enormous balls of gas squeezed so tightly that atoms crash together and fuse.

Fusion is the universe's favourite magic trick: turning tiny particles into bursts of pure energy. Our Sun? It's just one example.

A very convenient example, close enough to warm your face, power your plants, and give you freckles if you stay outside too long.

But it's not lonely. There are billions — yes, billions — more stars scattered around the galaxy, each with its own storms, planets, rhythms, and secrets.

The night sky is not quiet.

It's a glowing crowd.

STAR BABIES

Stars begin their lives in places that don't look very star-like at all: nebulae.

A nebula is a huge cloud made of dust, gas, and unfulfilled potential.

If you could scoop up just a teaspoon of nebula, it would look like smoke and feel like nothing.

But on a cosmic scale, a nebula holds enough material to make thousands of stars.

Inside these dusty nurseries, gravity is the bossy babysitter.

When a patch of the nebula becomes dense enough, gravity squeezes it tighter and tighter.

The gas heats, the pressure rises, and suddenly — flash! Fusion begins in the centre. A star is born.

Its first glow is like its first breath: a hot, bright "hello" to the universe.Planets sometimes form right beside it, from leftover crumbs of dusty material.

Imagine baking a cookie so enormous that pieces of dough fall off and become miniature worlds.

That's how solar systems begin: with cosmic baking accidents.

HOW THEY SHINE

Inside every star is a never-ending tug-of war. Gravity pulls inward, trying to squish the star flat.

Fusion pushes outward, blasting energy and saying, "No squishing today!"

This balance keeps a star stable for millions — often billions — of years.

Most stars fuse hydrogen into helium.

When two hydrogen atoms merge, a tiny bit of their mass transforms into energy, thanks to Einstein's famous equation:

$E = mc^2$, which basically means even a small amount of matter can unleash a ridiculous amount of energy.

That energy travels outward from the star's core, bouncing from particle to particle for ages before escaping as light.

Some of that light left the Sun eight minutes ago and is now warming your cheeks.

Your daily sunshine is a little parcel of nuclear peace sent across 150 million kilometres of emptiness.

THE COLOURS OF STARS

Not all stars are created equal.
Some glow red like a sleepy ember.
Some blaze blue like a brand-new flame.
Some shine white or yellow, like our Sun.
A star's colour tells us its temperature.
Blue stars are the hottest — so hot they burn through fuel like impatient sprinters.
Red stars are the coolest — calm marathoners pacing themselves for trillions of years.
Yellow stars sit comfortably in the middle, the cosmic equivalent of someone jogging with a pleasant playlist.
If stars were athletes:

- Blue stars would win every sprint but retire early.
- Red stars would jog forever.
- Our Sun would enjoy a steady morning run and a balanced breakfast afterward.

STAR FAMILIES

Stars rarely live alone.
Many come in pairs called binary systems — partners forever circling each other in gravitational dance.
Some are triples. Some are even quadruples.The universe loves complicated family trees.
Other stars form in massive clusters, thousands born from the same nebula.
They're stellar siblings, growing up together before drifting apart across the galaxy.
All these star families gather inside a galaxy — a glittering city of light.
The Milky Way, our home galaxy, holds more than 100 billion stars.
You live in one tiny suburb of this shining metropolis, tucked into a spiral arm called Orion.

WHEN STARS GROW OLD

Stars don’t stay youthful forever.

When a star runs out of hydrogen in its core, fusion slows down. Gravity pushes inward, the core shrinks, and the outer layers swell outward like a balloon.

The star becomes a red giant — enormous, cooler on the outside, burning brighter than ever.

One day, our Sun will expand so much that Mercury and Venus will vanish inside it.

Earth will get a dramatic, close-range suntan — the last one.

Bigger stars meet even more dramatic ends.

When their fuel runs out, their cores collapse in a violent instant.

Thc outer layers rebound in an explosion called a supernova — bright enough to outshine entire

galaxies for weeks.That blast scatters heavy elements into space: iron for blood, oxygen for breath, gold for jewellery, calcium for bones.

Almost everything that makes you came from a star that died in spectacular style.

STRANGE ENDINGS

When the fireworks settle, the leftovers are unusual.
Small stars end their days as white dwarfs — tiny, hot embers slowly cooling for trillions of years.
Medium stars collapse into neutron stars, made of matter crushed so tightly that a teaspoon of it weighs as much as a mountain.
And the largest stars become black holes, where gravity becomes unbeatable.
Not even light escapes.
A black hole is not a hole; it's the memory of a fallen star, folded into a point so dense that space itself curves around it.
The universe recycles everything, including gravity.

TA-DA!

READING THE LIGHT

Astronomers are cosmic detectives.Instead of fingerprints, they examine starlight.
When we pass starlight through a prism, it splits into a rainbow.
But it’s not just any rainbow: it has thin dark lines — missing colours.
Each element absorbs certain colours, leaving a pattern.
This pattern is like a barcode that tells us what the star is made of.
By reading these patterns, astronomers can learn a star’s age, temperature, motion, and even
whether it has planets tugging on it.
The universe glows in code, and we’re getting better at decoding it.

STARDUST AND YOU

Here's a cosmic truth that sounds like poetry:
You are made of stardust.
The iron in your blood, the calcium in your bones, the oxygen you breathe —
all were forged inside ancient stars.
When those stars exploded as supernovae, they scattered these ingredients into space.
Some of that dust drifted together, formed a nebula, and eventually became the Sun, Earth, and
everything on it.Life is what happens when stardust becomes curious

THE ENDLESS FIREWORKS

Some stars are being born right now in faraway nebulae.
Others are taking their final breaths in giant explosions.
The universe is a grand recycling program with excellent special effects.
Next time you look up, remember:
each sparkle is a story millions or billions of years long.
And you — a tiny, bright thought on a small blue planet — are part of the same story.

CHAPTER 2.
HOW OLD IS A STAR?

THE GREAT STAR WEB

Galaxies look like glowing pinwheels from afar. But each pinwheel is made of hundreds of billions of stars orbiting a mysterious centre.

Our galaxy spins like a cosmic merry-go-round, taking 230 million years for one full turn. If you rode the Milky Way like a carousel horse, dinosaurs would have come and gone before one lap finished.

The spiral arms are streams of stars flowing like rivers of light.

And we? We're drifting through one of them, passengers on a vast, spinning web woven by gravity.

THE STAR CLOCK

Stars don't have tick-tock clocks, but they tell time in their colours.
A blue star is young and furious.
A yellow one is middle-aged and calm.
A red one is elderly, glowing softly in its long twilight.
Star clusters show their ages the same way.
If a cluster has bright blue stars, it's young.
If it only has red ones, it's old.
Look up at night and you're witnessing all of cosmic history at once —
new stars being born, old stars fading, and light from every era arriving together in your eyes.

GIANTS AND DWARFS

If the Sun were the size of a basketball, the biggest stars would be stadiums.The smallest would be marbles.

Supergiants can swallow whole orbits.

Red dwarfs could hide inside Jupiter.

But small does not mean weak.

Red dwarfs live slow, steady lives, burning their fuel so carefully that they'll outlive every other star in the universe.

Trillions of years from now, most stars will be gone — but the red dwarfs will still be glowing like tiny candles in the dark.

SUPERNOVA SYMPHONY

A massive star’s life ends with a dangerous concert.
Its core collapses inward faster than you can blink.
The rebound creates a shockwave that tears the star apart from the inside.
For a few brilliant weeks, the supernova shines brighter than the galaxy it lives in.
Then its ashes scatter through space, carrying heavy elements — gold, silver, oxygen — to places where new stars and planets will form.
Every explosion is also an invitation:
“Here are the ingredients. Make something new.”

BLACK HOLES: THE SILENT HEARTS

What happens when gravity wins completely?
A star collapses into a point so dense that space-time folds.
This is a black hole — a region where the escape velocity is higher than the speed of light.
We cannot see black holes directly, but we see how nearby stars move.
They whirl faster, as if dancing around an invisible partner.
Black holes are cosmic monsters, yes — but without them, galaxies would lose their shape.
Even the universe needs a few mysterious supervisors.

NEBULA COMEBACKS

When a star explodes, it leaves behind a colourful cloud — a nebula.
The Crab Nebula, the Veil Nebula, and many others are the glowing ghosts of long-dead stars.
Inside these clouds, gravity begins its work once again.
Dust drifts. Gas gathers. Heat rises.
New stars take shape.If you could smell a nebula (you can’t — no air), it would smell like burnt metal and new beginnings.

OUR STAR, THE SUN

Our Sun is middle-aged — steady, bright, and sensible.
It spins every 27 days, powers every tree, breeze, wave, and heartbeat on Earth, and holds eight planets in orbit without complaining.
Inside its core, 600 million tons of hydrogen fuse into helium every second.
Even at that outrageous rate, the Sun still has about five billion years left.
One day it will swell into a red giant, then shrink into a white dwarf.
Its light will fade, but its influence will not be forgotten — after all, it raised every lifeform on Earth.

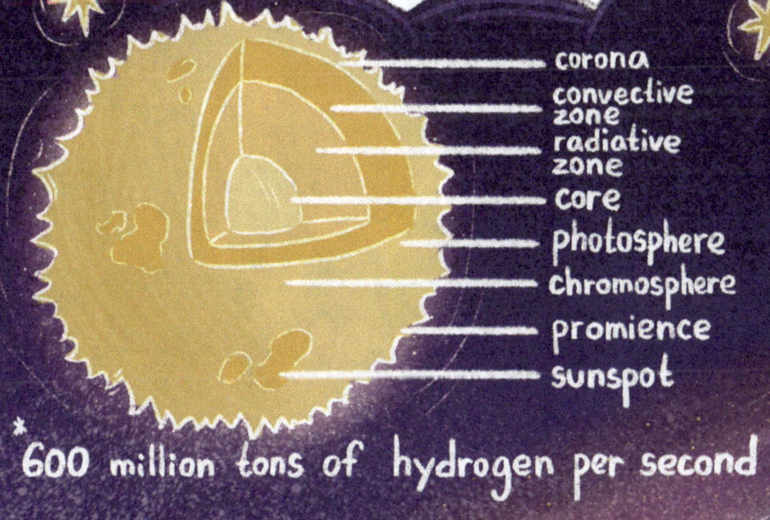

CONSTELLATIONS: STAR MAPS OF IMAGINATION

Humans have always drawn stories on the sky.
Orion the hunter, Scorpius the scorpion, the Southern Cross guiding travellers home.
Constellations aren't real shapes — just stars aligned from our perspective.
But these patterns helped early people travel, farm, tell stories, and understand time.Science and myth have always shared the same canvas.
Two ways of making sense of the same infinite stage.

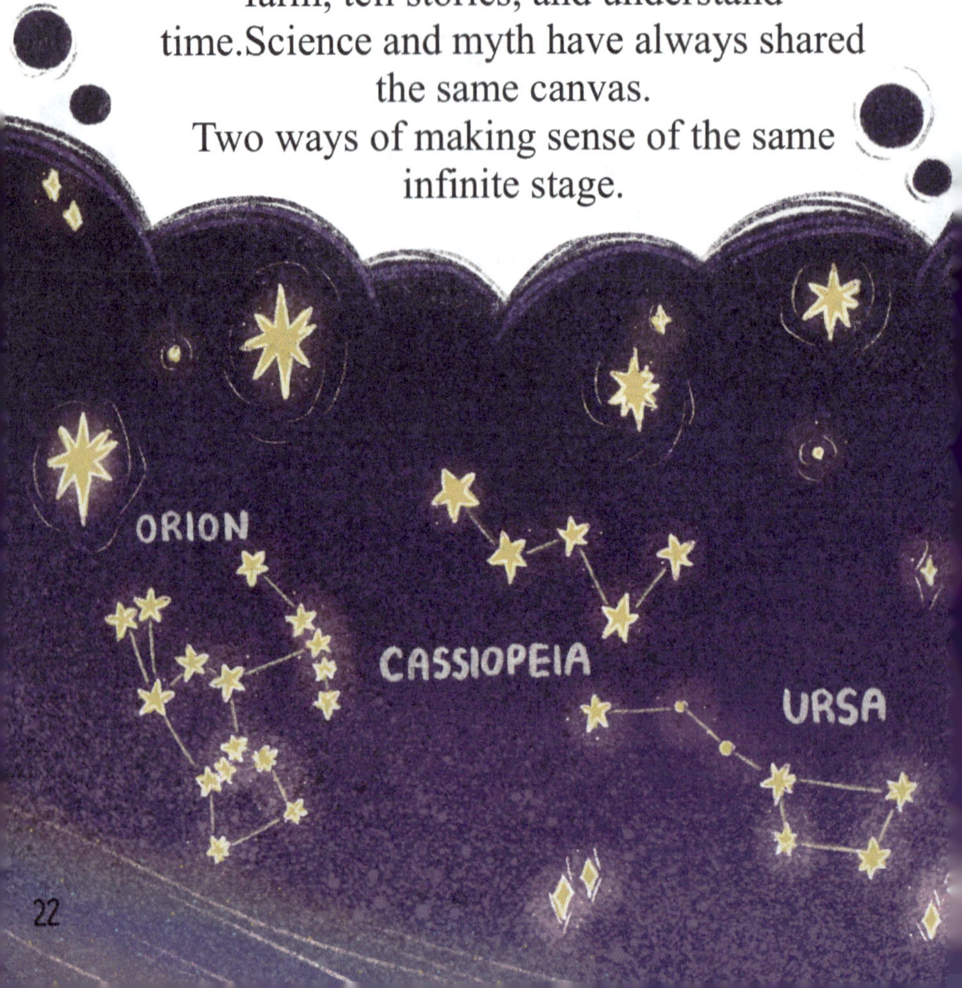

SEEING THE PAST

Light takes time to travel.
When you look at Alpha Centauri, you see it as it was four years ago.
Betelgeuse? 600 years ago.
Some galaxies are so far away that their light began its journey before humans existed.
The night sky is a time machine.
Every star you see is an old photograph glowing in the present.

YOU, THE LITTLE STAR

Your heartbeat echoes the steady rhythm of fusion.
Your atoms once shone inside a star, drifted through space, gathered in a cloud, and became you.
Your thoughts are stardust organizing itself into curiosity.
Next time you look up, whisper: “I see myself among the stars.”
And the stars, ancient and patient, whisper back:“Welcome home.”

CHAPTER 3.
GALAXIES: THE STAR CITIES

Our Cosmic Address

If the universe were a giant postal system, your address would look something like this:

You 

Planet Earth 

Solar System 

Spiral Arm Orion 

Galaxy: The Milky Way 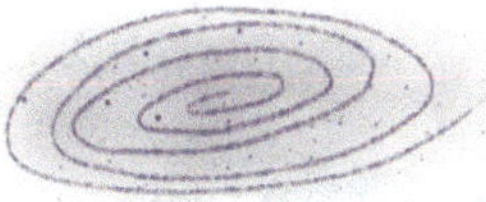

Universe 001 

Very dramatic.
Our galaxy is about 100,000 light-years wide — which is science's polite way of saying "absurdly huge."
We orbit the galactic centre at 220 kilometres per second.
It sounds fast, but you don't feel it because space is kind enough not to blow your hair around.

GALACTIC SHAPES

Galaxies come in shapes that would make a sculptor nervous.
Some are spirals, graceful and curling like cosmic seashells.
Some are elliptical, shaped like squashed footballs or smooth glowing blobs.
And then we have the irregular galaxies — the troublemakers.
"Irregular" is astronomy's way of saying "we have absolutely no idea what happened here, but it looks like someone dropped it."
Galaxies are messy, magnificent, and full of surprises — just like families.

NEIGHBOURHOOD WATCH

Our nearest big neighbour is the Andromeda Galaxy.
And it's headed straight for us.
Before you pack your emergency snacks: relax.
The collision won't happen for about four billion years.
Even when we finally meet, the stars won't smash together — they're too far apart.
It'll be the gentlest, most polite cosmic collision ever.
Two enormous cities of stars passing through each other, like ghosts shaking hands.

DARK MYSTERY

Galaxies behave as if they're much heavier than the matter we can see.
Stars orbit too fast.
Something unseen is holding them together.
This missing mass is called dark matter — invisible, untouchable, and possibly the universe's most mysterious introvert.
It doesn't glow, reflect, or interact with light at all.
But it makes up most of the universe, quietly pulling the strings from behind the scenes.
Whatever dark matter is, it's winning the cosmic popularity contest by sheer quantity.

CHAPTER 4.
THE LIFE AND DEATH OF LIGHT

LIGHT-YEARS EXPLAINED

A light-year isn't time; it's distance.
It's how far light travels in one year: about 9 trillion kilometres.
If you tried to drive that distance, you would need:

• several lifetimes,
• snacks for a dozen generations,
• and the world's most patient GPS saying, "Recalculating… please slow down… you've missed the turn… by 9000 billion kilometres."

Light doesn't wait.
It sets the speed limit for the entire universe.

STAR SPECTRA

When starlight passes through a prism, it spreads into a rainbow.
But look closer: thin dark lines slice the colours.
These lines are caused by atoms in the star's atmosphere absorbing certain wavelengths of light — like a barcode stamped on the rainbow.
Astronomers read these "rainbow barcodes" to learn what stars are made of.
Hydrogen, helium, oxygen — each leaves a distinctive mark.
The universe is constantly sending us coded messages.
Spectroscopy is how we read them.

THE COLOUR OF TIME

Blue stars burn fast and die young. Yellow stars like our Sun shine steadily.
Red stars glow softly for eons — cosmic grandparents with long stories to tell.
A star's colour is a time machine.
By looking at the sky, you're seeing stars at every stage of life at once.
It's like watching a movie in which all the scenes play together — and still somehow make perfect sense.

WHEN STARS WHISPER

Not every star shines steadily.
Some pulse, flicker, or change brightness over time.
These are variable stars — the universe's rhythmic storytellers.
Some throb like beating hearts.
Some flash like interstellar lighthouses.
Some dim and brighten because planets pass in front of them.
These whispers tell astronomers how big a star is, how fast it beats, and sometimes even if it has company.
The cosmos loves rhythm.

CHAPTER 5.
THE EXPLODING UNIVERSE

SUPERNOVA FIREWORKS

When a giant star reaches its end, it does not go quietly.
It collapses inward, rebounds outward, and explodes in a spectacular supernova.
For a brief moment, one dying star can shine brighter than billions of other stars combined.
It is the universe shouting “TA-DA!” with all its remaining strength.
If you ever feel insignificant, remember this: every atom in your body was once part of a cosmic fireworks show.

THE HEAVY METAL FACTORY

Supernovae create the heavy elements.
Iron for your blood.
Calcium for your bones.
Gold for rings.
Silver for cutlery.
Uranium for… well… scientists to argue about. Without exploding stars, the universe would contain no gold, no oxygen, and definitely no humans.
Dinner parties, jewellery stores, and biology exams all exist because stars blew themselves apart.

REMNANTS AND GHOSTS

After a star explodes, a cloud of glowing debris remains — a supernova remnant.

Some have poetic names:

- the Crab Nebula
- the Veil Nebula

Others have less poetic names like "G292.0+1.8," which is what happens when scientists run out of imagination but still need labels.

These remnants hum with leftover energy, lighting the sky like cosmic campfires.

THE COSMIC RECYCLE BIN

Nothing in the universe is wasted.
Supernova debris eventually drifts into clouds, gathers, collapses, and forms new stars.
Planetary systems grow from the leftovers.
Life eventually flourishes on tiny rocky worlds.And in time, those planets produce curious beings who look up and ask questions.
Destruction becomes creation, again and again.
The universe is the ultimate recycler.

CHAPTER 6.
BLACK HOLES AND STRANGER THINGS

WHEN GRAVITY WINS

Imagine squeezing a star into something the size of a sugar cube.
That level of density is beyond imagination — but the universe did it anyway.
A black hole forms when gravity becomes so strong that not even light can escape.
It isn't a hole; it's a region where physics takes a deep breath and says, "Well… this just got interesting."

EVENT HORIZON ETIQUETTE

The “edge” of a black hole is called the event horizon — the point of no return.

If you fall in, you don’t notice anything unusual when crossing the edge.

But to an outside observer, you appear to stretch like spaghetti and freeze in place. This process is called spaghettification, a word that makes Italian astronomers sigh heavily.

SUPERMASSIVE SURPRISES

At the centre of most galaxies lurk supermassive black holes, millions or billions of times heavier than the Sun.

Our own galaxy has one called Sagittarius A*.
Astronomers added the asterisk because, frankly, the thing deserves dramatic punctuation.
These giants don’t swallow galaxies whole — instead, they help keep everything in orbit, like gravitational anchors.

HAWKING'S WHISPER

Black holes aren't entirely silent.
They slowly leak energy through something called Hawking radiation, named after Stephen Hawking.
It means black holes can shrink and eventually evaporate — extremely slowly.
Even the darkest mysteries fade in time.
Nothing lasts forever, not even shadow.

CHAPTER 7.
STAR TOOLS

THE TELESCOPE REVOLUTION

Telescopes are humanity's superpower. From Galileo's tiny spyglass to modern giants like Hubble and the James Webb Space Telescope, each upgrade extends our vision deeper into the cosmos. Telescopes turn dim smudges into galaxies, and galaxies into entire histories. Every time we build a better one, we basically shout, "Enhance!" — and the universe obliges.

RADIO EARS

Not all telescopes use light.
Some “listen” to the universe using radio waves.
Radio telescopes catch the whispers of quasars, pulsars, and even the faint hum of the early universe.
If there were cosmic FM stations, radio astronomers would be the first to tune in.

INFRARED GOGGLES

Infrared telescopes can see heat — even the warmth of dusty clouds where stars are born. They peer through darkness the way night-vision goggles reveal wildlife in the forest. Infrared astronomy lets us watch stars in their earliest, messiest, sleepiest stages — cosmic baby monitors.

WHY WE KEEP LOOKING

Every telescope ever built has one purpose:
curiosity.
We look because we're human.
We wonder because we exist.
And the moment we stop asking questions,
we stop growing.
The universe never runs out of mysteries.
Luckily, curiosity never runs out either.

CHAPTER 8.
OUR SUN: THE FRIENDLY STAR

SOLAR PERSONALITY

The Sun is polite but powerful.
Polite, because it sends warmth and light
across space instead of blasting
us with fireballs.
Powerful, because without it, Earth would
become a sad, dark marble
floating in eternal winter.
Sometimes the Sun throws tantrums — solar
flares — that send charged particles zipping
toward Earth.
Most of them miss.
The few that hit just give us pretty auroras
and very grumpy satellite engineers.

SUNSPOTS

Sunspots are cooler, darker patches caused by magnetic activity.
They come and go in an eleven-year cycle.
During the "solar maximum," the Sun looks like it has cosmic acne.
During minimum, it looks freshly washed.
Sunspots can grow larger than planets, reminding us that even stars have mood swings.

AURORAS

When solar particles hit Earth's magnetic field, they slide toward the poles and collide with our atmosphere.The sky glows green, purple, and red — auroras dancing like floating curtains of light.

The North and South Poles get the best shows.

Penguins and polar bears live far apart, but both have front-row seats to nature's greatest screensaver.

SOLAR FAREWELL

In five billion years, the Sun will expand into a red giant.
It will swallow Mercury and Venus and give Earth a dramatic final roast.
Eventually it will shed its outer layers, leaving behind a white dwarf — a tiny ember glowing with ancient pride.
By then, humanity will hopefully have found new homes around other stars.
No rush — five billion years is plenty of time to pack.

CHAPTER 9.
PLANETS AND THEIR STORIES

THE LEFTOVERS OF CREATION

When stars form, not all the dust and gas falls into them.
Some remains swirling around like tiny cosmic crumbs.These crumbs bump, stick, clump, and grow into planets.
Every world — rocky or gassy — is a piece of star-making rubble that refused to burn.
Planets are the rebels of creation.

THE ROCKY QUARTET

Mercury, Venus, Earth, and Mars are the rocky planets — small, tough, and built from metal and stone.
Earth is the only one with liquid water, breathable air, and life.
A miracle of chemistry and just the right distance from the Sun.
Mars once had rivers, lakes, and maybe even oceans.
Now it's a frozen desert with giant volcanoes and dust storms.
The universe loves plot twists.

THE GAS GIANTS

Jupiter and Saturn are enormous gas giants. Mostly hydrogen and helium — they're almost stars, but not quite.Jupiter is home to the Great Red Spot, a storm so big it could swallow Earth several times.
And it's been raging for centuries, which is impressive dedication for a weather system.
Saturn's rings are made of ice and rock, shining like a cosmic jewellery set that the universe wears on special occasions.

THE ICE OUTSIDERS

Uranus and Neptune are called ice giants — cold, windy, distant worlds.
Uranus spins on its side, rolling around the Sun like a sleepy barrel.
Neptune has winds faster than sound, howling across its dark-blue atmosphere.
Beyond them drift comets, asteroids, and icy leftovers — pieces of the early solar system waiting for someone curious enough to visit.

CHAPTER 10.
MOONS: THE BORROWED LIGHTS

EARTH'S COMPANION

The Moon is Earth's first friend.
Silent, cratered, and steady — always showing the same face.Its gravity pulls on our oceans, creating tides that help life flourish.
It might even tug slightly on our moods, though scientists are still debating that part.
Without the Moon, Earth's nights would be darker and our oceans far less dramatic.

THE MANY MOONS OF OTHERS

Other planets have more moons than they know what to do with. Jupiter has over ninety.

Some are icy worlds hiding oceans deeper than Earth's.

Europa may have warm water beneath its frozen shell — a secret ocean lit only by chemistry.

Io is the opposite: a volcanic troublemaker covered in erupting lava fountains.

Geysers, flames, explosions — Io is basically a rock with anger-management issues.

Saturn's moon Titan has lakes of liquid methane — imagine navigating a sea of natural gas under an orange sky.

The solar system is full of strange, wonderful neighbourhoods.

MOONLIGHT MYSTERIES

Moons form in all sorts of chaotic ways.
Some are born from collisions.
Some are captured asteroids.
Some break apart and reform.
Each moon carries a story of violence turned into harmony.
The solar system is peaceful now, but it was built out of chaos.

LUNAR FAREWELL

The Moon is drifting away from Earth — about four centimetres a year.
In a billion years, it will look smaller in the sky.
Eclipses will be less dramatic.
But the tides will still remember the Moon's ancient pull, long after it has floated to a safer distance.

CHAPTER 11.
THE COSMIC OCEAN

SAILING THE VOID

Space isn't empty.It's filled with dust, radiation, particles, and energy.
But it's thin — incredibly thin — like an ocean where the water molecules are shy and rarely meet.
Spacecraft like Voyager drift through this vast sea, carrying human greetings, music, and golden records into the unknown.
The universe listens, quietly.

COSMIC BACKGROUND

The faint microwave glow across the entire sky is the cosmic microwave background — the oldest light we can see.
It's a baby photo of the universe when it was only 380,000 years old.
Tiny ripples in this ancient glow show where galaxies would one day form.
Every pixel is a fossil.
Not of bones — but of time.

DARK ENERGY'S DRIFT

The universe is expanding faster and faster. Something invisible is pushing galaxies apart. We call it dark energy, which is science's elegant way of saying, "We have no idea, but it's definitely doing something." Dark energy is the ultimate mischief-maker, stretching space like cosmic taffy.

THE EDGE OF EVERYTHING

The universe has no edge.
Space curves, stretches, expands.
Travel far enough in a straight line and, in theory, you might eventually return from behind yourself.
Infinity has a sense of humour.

CHAPTER 12.
ALIENS, LIFE, AND LITTLE BLUE DOTS

THE GOLDILOCKS ZONE

Life likes things just right: not too hot, not too cold — the Goldilocks zone.
Earth sits perfectly in its star's comfortable warmth.
Not scorched. Not frozen.Just right for lakes, rivers, clouds, and poetry.
Scientists search other stars for planets in similar cosy orbits.
We've already found thousands.
The universe might be sprinkled with other "just-right" worlds.

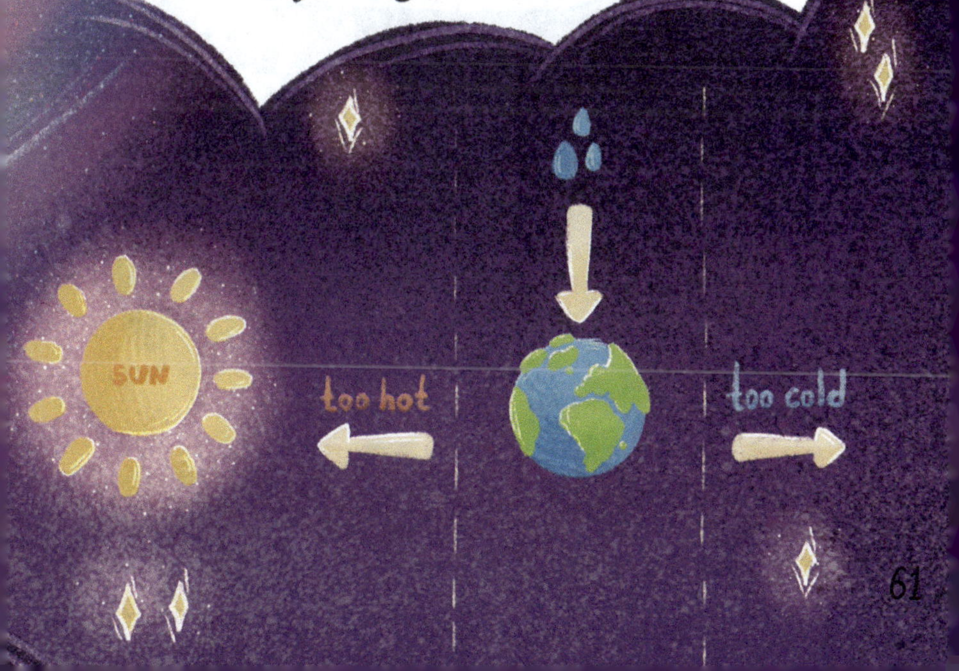

THE INGREDIENTS OF LIFE

Life's recipe is surprisingly simple:
Carbon, hydrogen, nitrogen, oxygen,
phosphorus, and sulphur.
Add energy, chemistry, and time — lots of
time — and life bubbles into being.
Maybe the universe has cooked this recipe
more than once.
Maybe life is not rare, just patient.

MESSAGES IN THE DARK

For about a century, Earth has been leaking radio signals into space — old broadcasts, music, news, sitcoms, and occasional static. If someone out there is listening, our first message is already on the way.
It just depends on where "out there" is.Space is big. It might take a while.

ALONE OR TOGETHER?

Silence doesn't always mean loneliness.
It might mean distance.
Or difference.
Or simply that civilizations don't
communicate the way we expect.
The search continues because curiosity
refuses to sit still.
Life, like light, keeps travelling.

CHAPTER 13.
TIME MACHINES AND COSMIC SPEED LIMITS

THE SPEED OF LIGHT

Light travels at 299,792 kilometres per second — the universe's ultimate speed limit. Nothing with mass can go faster. If you try, physics simply crosses its arms and says, "Non."The faster you move, the slower your time passes — a cosmic joke Einstein pointed out, and the universe found it so funny it made it a law.

TIME DILATION

For astronauts travelling close to light speed, time really does slow down.
A high-speed space trip could make you return younger than your friends.
Relativity proves that fairness is relative and time is negotiable.

WORMHOLES

Imagine bending space like a sheet of paper
and poking a shortcut through.
That shortcut is a wormhole.
The maths says they might exist.
The engineering says, "Oui, but… not yet."
Imagination travels faster than physics
allows.

THE FUTURE LIGHT CONE

Everything that can affect you lies inside
your light cone — the region where signals
can reach.
Beyond it lies elsewhere — unreachable,
unknowable.
Even infinity has boundaries drawn by
geometry.

CHAPTER 14.
HOW TO WEIGH THE UNIVERSE

COSMIC SCALES

How do you weigh something as huge as a galaxy?
You don't put it on a scale, obviously.
Instead, you watch how fast stars orbit.
If they're moving faster than expected, something must be adding extra gravity.
That something is usually dark matter, hiding in plain… nothing.

MEASURING LIGHT

To measure distance in space, astronomers use standard candles — stars whose brightness is predictable. By comparing how bright they look from Earth to how bright they should be, we calculate distance. It's the universe's built-in measuring tape.

THE EXPANDING UNIVERSE

In 1929, Edwin Hubble discovered that galaxies are moving away from us — and the farther they are, the faster they flee.
Space itself is stretching.
Galaxies aren't flying outward; the fabric of the universe is expanding beneath them.
The cosmos grows like rising dough — puffing outward everywhere at once.

COSMIC BACKGROUND REVISITED

Tiny ripples in the cosmic microwave background show where matter clumped together in the early universe.
Those ripples became galaxies, stars, planets, and eventually… beings capable of looking back and studying them.
You are literally reading the universe's earliest handwriting.

CHAPTER 15.
THE FATE OF EVERYTHING

THE GREAT FREEZE

If expansion continues forever, galaxies will drift apart, stars will fade, and everything will cool into darkness.
Energy spreads out.
Entropy rises.
The universe becomes a cold, quiet place.
This is the heat death — the final winter.

THE GREAT CRUNCH

But maybe gravity will fight back.
Maybe the universe will slow, stop, and collapse inward.
Everything would rush together in a final crunch — a cosmic inhale after the long exhale.
Another Big Bang could follow.
A new universe.
A fresh beginning.

THE GREAT RIP

Or dark energy might grow stronger and
stronger…
Stretching galaxies apart…
Then solar systems…
Then atoms…
Until even space-time snaps.
This is the Big Rip, the universe's most
dramatic breakup.
Let's hope dark energy isn't that ambitious.

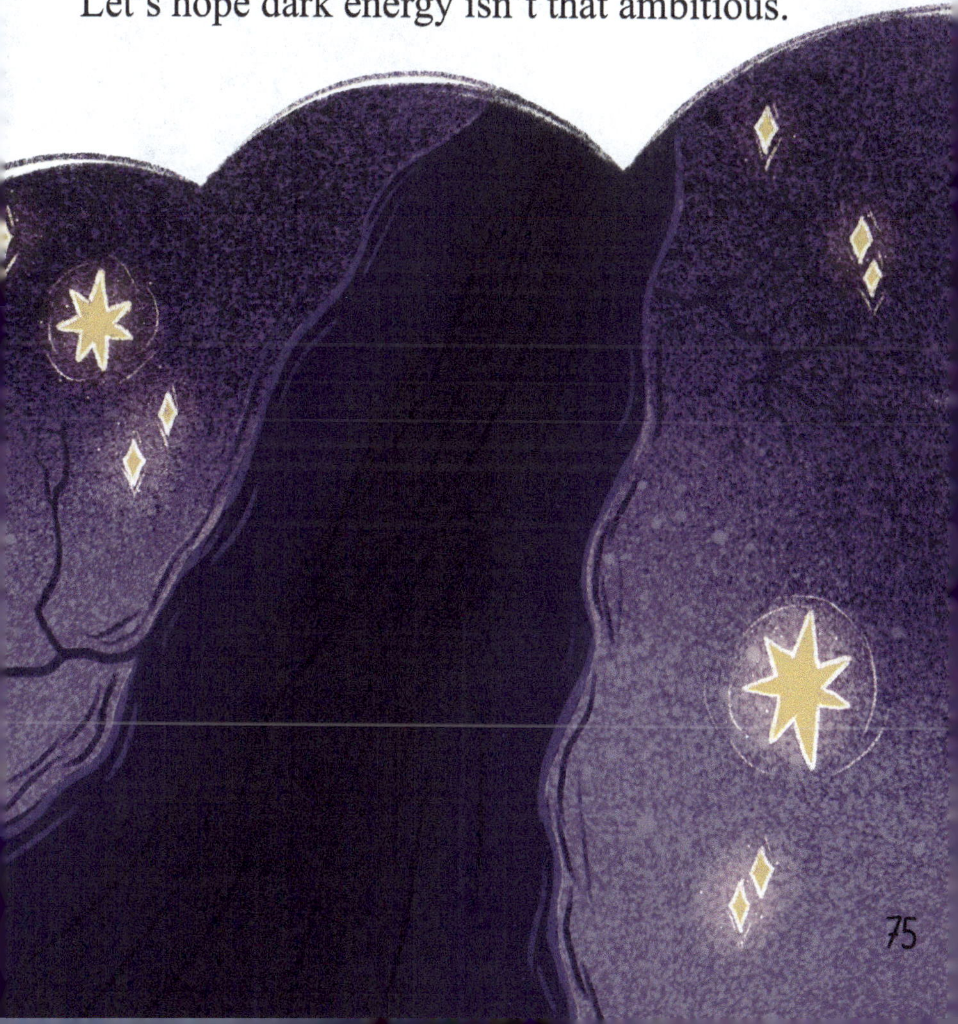

CYCLES OF WONDER

Whatever fate awaits, energy never disappears.
It changes forms, dances new dances, sparks new stories.
Questions remain.
Mystery persists.
And wonder grows.
New minds will ask old puzzles again. The universe never stops inviting curiosity.

CHAPTER 16.
THE LANGUAGE OF THE UNIVERSE

MATH AS MUSIC

Mathematics is the universe's sheet music. Equations are melodies that explain how stars shine, how planets move, and how galaxies swirl.
Einstein's $E = mc^2$ is a simple tune with explosive consequences.
Math is the quiet instrument behind all cosmic performances.

PATTERNS EVERYWHERE

Nature loves patterns.
Spiral galaxies resemble seashells.
Lightning forks like brain neurons.
River deltas mimic tree branches.
The universe reuses its favourite designs like an artist who refuses to clean out their brush jar.

SYMMETRY AND BEAUTY

Physicists chase symmetry because it reveals hidden rules.
When the maths fits perfectly, the universe seems to sigh in satisfaction.
Beauty isn't just in the stars — it's in the equations describing them.

CHAOS AND ORDER

Even chaos has rhythm.
Storms form patterns.
Galaxies swirl.
Hearts beat.
Planets orbit.
Order and disorder dance through everything
— a cosmic duet.

CHAPTER 17.
COSMIC STORYTELLERS

ANCIENT EYES

Long before telescopes, humans watched the sky and told stories.Constellations were calendars, clocks, and campfire tales. The sky was a map of time.

SCIENCE AND MYTH

Myth said Orion hunted across the heavens.
Science says he's a cluster of hot young stars burning hydrogen.
Two different truths — one poetic, one scientific — sharing the same sky.

ARTISTS OF INFINITY

Every painting of the night sky is a record of feeling.
Art preserves wonder the way science preserves facts.
Both are ways of seeing the universe.

THE UNIVERSE WITHIN

We study stars not to escape Earth, but to understand the light inside us. Curiosity is a flame carried from ancient stargazers to modern astronomers.The universe expresses itself through our questions.

CHAPTER 18.
SPACE EXPLORERS

FIRST STEPS

Sputnik beeped.
Gagarin smiled.
Armstrong stepped.
Each small moment echoed across the vacuum, telling the universe that humans had learned to leave home.

THE ROBOT SCOUTS

Robots have gone where humans can't.
Rovers crawl across Mars.
Probes fly past gas giants.
Space telescopes orbit Earth like electronic eagles.
They are our metal messengers, brave and tireless.

THE INTERNATIONAL CREW

Space belongs to everyone.
Astronauts come from different countries and speak many languages, but they share the same mission: exploration.
Above Earth, borders vanish and curiosity becomes the only passport.

THE NEXT FRONTIER

One day, we may build homes around other stars.
But exploration isn't escape — it's expansion of belonging.
The universe is vast, and we're only just beginning to say bonjour.

CHAPTER 19.
COSMIC CATASTROPHES

ASTEROID VISITORS

Now and then, rocks fall from the sky.
Some tiny, some enormous.
They've ended species and reshaped continents.
But they've also delivered water, organics, and surprises.Asteroids are both villains and heroes in Earth's history.

GAMMA-RAY BURSTS

The most powerful explosions known are gamma-ray bursts — jets of energy from dying stars or colliding neutron stars.
A single burst could sterilize an entire galaxy.
Fortunately, none are pointing at us.
Probably.
(Science always leaves a little room for "probably.")

COLLIDING STARS

When two neutron stars spiral together and collide, they stretch space itself. These ripples are gravitational waves — the universe ringing like a cosmic bell. We can now detect these waves, listening to the cosmos through the fabric of reality.

THE FRAGILE MIRACLE

For all the chaos — explosions, collisions, radiation — life persists.
It grows in oceans, clings to rock, hides in ice, and thrives in sunlight.
Life is the universe's most delicate rebellion.

CHAPTER 20.
THE FUTURE SKY

NEW EYES ON THE UNIVERSE

Future telescopes will see the first stars ever born — the universe's opening act. We're about to watch creation replay its earliest memories.

TERRAFORMING DREAMS

Mars may one day have seas again.
But before we reshape other worlds, we must learn responsibility — how to care, not just conquer.
Exploration demands humility.

POST-HUMAN HORIZONS

Perhaps AI or something even stranger will continue our curiosity across the galaxy. The universe doesn't care who asks the questions.
Only that the questions continue.

THE FINAL SPARK

When the last star fades, new eyes — whoever they belong to — will find ways to see in the dark.
Stories will continue, because stories, like light, never truly end.
And you, little stardust, are part of that eternal glow.

Made in United States
Troutdale, OR
11/03/2026

56845373R00056